Raphaela A. Duarte Silveira

Difficulties in marine animal rehabilitation centres

Raphaela A. Duarte Silveira

Difficulties in marine animal rehabilitation centres

ScienciaScripts

Cover image: www.ingimage.com

This book is a translation from the original published under ISBN 978-620-2-04074-7.

Publisher:
Sciencia Scripts
is a trademark of
Dodo Books Indian Ocean Ltd. and OmniScriptum S.R.L publishing group

120 High Road, East Finchley, London, N2 9ED, United Kingdom
Str. Armeneasca 28/1, office 1, Chisinau MD-2012, Republic of Moldova, Europe
Managing Directors: Ieva Konstantinova, Victoria Ursu
info@omniscriptum.com

Printed at: see last page
ISBN: 978-620-8-40970-8

ACKNOWLEDGMENTS

To my parents, Claudia and Olimpio, for their total support, dedication and interest in my academic life. Thank you for being by my side, encouraging me to always do my best. Without you, I wouldn't be the person I am today and I wouldn't have made it this far.

To all the rest of my family, but especially to my sister, Maria Fernanda. You always saw me as an example to follow, which motivated me to do my best every time.

To the biology boys and girls, Isadora, Larissa, Livia, Lorena, Natalia, Patricia, Thaisa and Clesio. You are more than my friends, you have become part of my family. Thank you for all these five years together, none of this would have been the same without you. I'll always remember our parties and drinking binges, our desperation before exams, my photocopied summaries, our post-test discussions to see if the answers were the same. More than that, you made my years at university some of the best of my life. I'll never forget everything we went through and you'll always have a little place in my heart. I love you so much!

My friend, Aline, who put up with me sharing the same room for three years. A companion for all times, whether for an open bar or an experience in the land of Uncle Sam. "Beuzinha", thank you for your friendship, understanding and affection. May this be one more stage of many for us.

To my friends at Ciencias sem Fronteiras, especially Carolina, Julio and Monica. You were very important to me and became my second family in a country I didn't know. With you by my side, I was able to feel a taste of home. Thank you for all our trips and parties, joys and sorrows. Without a doubt, this year has been unforgettable!

To my advisor, Dr. Carlos Saad, who, even though he hadn't known me for very long, agreed to help me during this final stage of my degree. Thank you for your patience, goodwill and willingness to help me whenever I needed it.

To my former advisor, Dr. Vania Techio, who taught me a lot about what I know today. Thank you for accepting me into your laboratory all those years, but above all, for encouraging me to follow the field I chose.

To my ever-honorable partner, Laiane, who was always willing to teach me something new. Thank you also for supporting me in my decisions and encouraging me to pursue what I wanted.

To the team at the Gremar Institute, who gave me my first contact with Magellanic penguins and made me love marine biology even more. Thank you for your attention, patience and willingness to teach me.

The Federal University of Lavras and the teachers I had throughout my degree.

Finally, I thank God for all the people I've met over the years, for all the experiences I've had and for the strength and blessing he's given me to get this far.

To everyone mentioned here, thank you from the bottom of my heart!!!

SUMMARY

Marine animal rehabilitation centres are undertakings authorized by the Brazilian Institute for the Environment and Renewable Natural Resources - IBAMA - with the main objective of receiving, sorting and rehabilitating marine species so that they can be reintroduced into their natural *habitat.* Aquatic species, especially migratory ones, suffer intense environmental pressures caused by anthropogenic damage to their breeding and feeding grounds. Species are often found stranded on beaches in a critical state of health or even dead, mainly due to ocean pollution. Therefore, rehabilitation centers are important for the recovery of these stranded animals, giving them a chance to recover and return to their habitat. For any species, successful rehabilitation depends on correct management as well as a location with suitable environmental conditions, which in turn depend on the biology, ecology, physiology and behavior of each species. There are reports of several marine species that have been rehabilitated and reintroduced into their natural habitat, such as chelonians (sea turtles), cetaceans (dolphins and whales), pinnipeds (sea lions, elephant seals, seals) and seabirds (gulls, albatrosses, boobies, penguins). *Magellanic* penguins *(Spheniscus magellanicus)* are the most common migratory seabirds in rehabilitation centers on the southwest Atlantic coast, migrating long distances in search of food in the warmer waters of the southern hemisphere. During their migration, many juveniles end up lost in the Falklands current and strand on the Brazilian coast. When recovering, penguins should be kept in places similar to their natural habitat, in order to facilitate and optimize their recovery, as well as avoiding injuries and diseases. Viral (flavivirus, for example), bacterial (entente, respiratory infections) and fungal (aspergillosis) diseases, as well as parasites, can complicate penguin recovery and even cause death. Most of the time, these diseases are contracted in the facilities of rehabilitation centers and the way to prevent this is to ensure that the enclosures are cleaned and disinfected on a daily basis. Therefore, this study aimed to describe the main shortcomings found in rehabilitation centers related to environmental conditions and the treatment of diseases contracted by penguins, especially Magellanic penguins, illustrating this with case reports of three juvenile penguins from the Gremar Institute - Research, Education and Wildlife Management.

Keywords: Aspergillosis. Pododermatitis. Penguins. *Spheniscus*

SUMMARY

1 INTRODUCED

Penguins belong to the *Spheniscidae* family, which contains 18 species distributed throughout the southern hemisphere. Although they don't breed in Brazil, four species have been recorded on the Brazilian coast: the king penguin *(Aptenodytes patagonicus* Millher, 1778), the yellow-fronted penguin *(Eudyptes chrysocome* Forster, 1781), the yellow-fronted penguin *(Eudyptes chrysolophus* Brandt, 1837) and the most common of them, the *Magellanic* penguin *(Spheniscus magellanicus).* The Magellanic penguin is a migratory species, meaning that during the winter in the southern hemisphere, the animals follow the Falklands chain in search of their main food, the anchovy *(Engraulis anchoita* Hubbs & Marini, 1935) and arrive on the Brazilian coast. However, many penguins end up stranded on Brazilian beaches during this process (ICMBIO, 2010).

The main factors responsible for penguins stranding on the Brazilian coast are: *habitat* destruction, disorderly tourism, lack of food due to overfishing, marine pollution including oil spills, incidental capture during fishing and several others (ICMBIO, 2010). Magellanic penguins are important components of the food webs of the ecosystems where they live, as they control the birth rate of several species and serve as food for others (SANDER, 2011). They are also environmental indicators, as the fragile state of conservation of their population reflects the conditions of the Southern Oceans (BOERSMA et al., 2011).

There are several marine animal rehabilitation centers throughout the southern hemisphere with the aim of receiving, screening and rehabilitating not only Magellanic penguins, but also other debilitated animals. Once the penguins arrive at a rehabilitation center, three fates are possible: a) death due to critical condition or euthanasia due to factors that make it impossible for them to be reintroduced into the environment; b) remaining in captivity due to their unsuitability for release, where they can be used for environmental education purposes; c) reintroduction into their natural habitat after successful rehabilitation. Rehabilitation centers need proper management, specialized staff and an enclosure structure with adequate environmental conditions to be able to provide a successful recovery for penguins. However, this is often not possible (BOERSMA et al., 2011; GEEVERGHESE, 2013).

The problems the centers encounter can be political, financial and biological. Often, the loss of genetic diversity in penguin populations is associated with a lack of information, financial resources and political support so that conservation actions can reflect a significant change in the *status of* this species (ICMBIO, 2010). In addition, the centers face problems related to the acquisition of diseases by the Magellanic penguins. The diseases that limit the recovery of these birds can be viral, bacterial, parasitic and especially fungal, such as aspergillosis. These diseases are associated with local pathogens or the stressful conditions of their environment (BOCARDO et al., 2009). The aim of this study was to describe the main difficulties encountered in rehabilitation centers related to environmental conditions and the treatment of diseases contracted by penguins, especially Magellanic penguins, illustrating this with case reports of three juvenile penguins from the Gremar Institute - Research, Education and Wildlife Management.

2 THEORETICAL FRAMEWORK

2.1 Taxonomy, Characteristics and Distribution of Penguins

Penguins, belonging to the order *Sphenisciformes,* make up a single family *Spheniscidae* with 18 species distributed throughout the southern hemisphere. The distribution of species in South America ranges from Tierra del Fuego to Peru and the Galapagos Islands in Ecuador. Of the seven species found in South America, only the *Magellanic* Penguin *(S. magellanicus* Forster, 1781) and the Humboldt Penguin *(Spheniscus humboldti* Meyen, 1834) have breeding colonies in the extreme south of the continent, while the other species have their colonies on continental islands (ICMBIO, 2010; SILVA FILHO & RUOPPOLO, 2014).

Penguins are considered to be the vertebrates best adapted to swimming, both in terms of speed and depth, as well as distance. Their anatomy favors their lifestyle, with adaptations to both aquatic and terrestrial environments. Although they cannot fly, their hydrodynamic, fusiform bodies with short necks and limbs and feet with interdigital membranes allow them to move quickly in the water. The dense, solid, non-pneumatic bones and wings modified into fins allow for long dives. Most of a penguin's life is marine, but like all birds, they return to land to breed (WALKER & BOERSMA, 2003; AZA, 2005; SILVA FILHO & RUOPPOLO, 2014).

All penguins have a white ventral part and a black or grayish dorsal part, with the plumage of the head varying mainly (Figure 1). Like all birds, they are endothermic, with a body temperature ranging from 37.8 to 38.9°C. Temperature regulation occurs due to physiological and behavioral adaptations. Overlapping feathers and a thick layer of fat help to insulate these birds. However, penguins found in areas with higher temperatures can have problems with excess heat. In these cases, they generally have thinner layers of fat and a lower density of feathers on the head and flippers. Other mechanisms that allow them to lose heat are keeping their flippers away from the body or staying in shelters in the substrate (AZA, 2005; SILVA FILHO & RUOPPOLO, 2014).

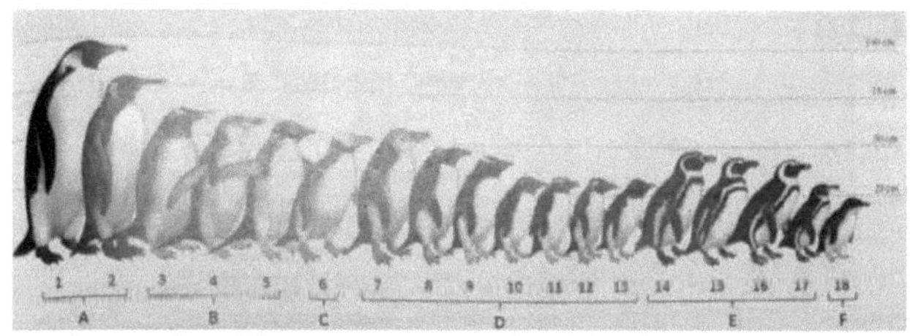

Figure 1 Penguin species. (1) Emperor penguin, (2) King penguin, (3) Pua penguin, (4) Antarctic penguin, (5) Adelie penguin, (6) Yellow-eyed penguin, (7) Royal penguin, (8) Macaroni penguin, (9) Erect-crested penguin, (10) Snares penguin, (11) Fiordland Penguin, (12) Northern Yellow-breasted Penguin, (13) Southern Yellow-breasted Penguin, (14) Humboldt Penguin, (15) Magellanic Penguin, (16) African Penguin, (17) Galapagos Penguin, (18) Blue Penguin. Genera (A) *Aptenodytes,* (B) *Pygoscelis,* (C) *Megadyptes,* (D) *Eudyptes,* (E) *Spheniscus,* (F) *Eudyptula.* Source: Silva Filho & Ruoppolo (2014).

Penguins are pelagic animals and can feed on fish, cephalopods and crustaceans depending on the species. Consumption can vary according to the animal's needs, but it is estimated that the amount is 2% to 3% of body mass in large species, and 10% to 20% in small species. (ICMBIO, 2010; SILVA FILHO & RUOPPOLO, 2014). Small species need a large amount of food because heat loss and energy expenditure to maintain thermoregulation in these animals are greater than in large species (SIWANOWICK, 2010). This body mass increases in the period leading up to moulting, a phenomenon in which penguins change all their feathers at once. Moulting causes penguins to lose their waterproofing, which restricts them to land for 2 to 4 weeks, fasting. This process requires a lot of energy, which makes these animals prepare by eating more than usual, and they can achieve an increase of 40% to 70% of their body mass in the form of fat, which will be lost during the process due to fasting (SILVA FILHO & RUOPPOLO, 2014).

Four penguin species have been recorded in Brazil, the most common being the *Magellanic* penguin *(Spheniscus magellanicus).* The other species that eventually migrate to the Brazilian coast are the King Penguin *(Aptenodytes patagonicus), the* Southern Yellow-breasted Penguin *(Eudyptes chrysocome)* and the Macaroni Penguin *(Eudyptes chrysolophus).* Although penguins have been recorded in Brazil, they do not breed in this area. During the winter months in the southern hemisphere, they migrate from their breeding grounds in search of food in warmer waters. However, many get lost in the Falklands current and end up stranding on the Brazilian coast (ICMBIO, 2010; GEEVERGHESE, 2013; SILVA FILHO & RUOPPOLO, 2014; NOGUEIRA et al, 2015).

2.2 Magellanic penguin

The Magellanic penguin, as well as three other species, the Humboldt penguin *(Spheniscus humboldti* Meyen, 1834), the African penguin *(Spheniscus demersus* Linnaeus, 1758) and the Galapagos penguin *(Spheniscus mendiculus* Sundevall, 1871) belong to the *Spheniscus* genus. The characteristics of this genus are the black and white pattern of the head, a black stripe along the animal's flank and exposed skin on the head (PENGUINWORLD, 2016).

Magellanic penguins can reach 65 cm in length and an average weight of 4 to

5 kilos as adults, with males generally larger than females (WILLIAMS, 1995; SICK, 1997). Adults of this species have a black head, beak and back, a white ventral area with a black stripe around the pectoral area and a white stripe that starts at the eyes, circling the face up to the throat, distinguishing these animals from other species of the genus. Young individuals are grayish and do not have a well-defined pectoral stripe (WILLIAMS, 1995; ICMBIO, 2010).

The distribution of the Magellanic penguin ranges from the island of Algarrobo in Chile to the Valdes Peninsula off the coast of Argentina, passing Cape Horn at the southern tip of the continent, as well as the Falklands Islands (Malvinas Islands). The colonies of San Lorenzo and Punta Tombo in Argentina are the most numerous (Figure 2). According to the International Union for Conservation of Nature - IUCN (2016), the Magellanic penguin population is estimated at 1.3 million pairs (950,000 in Argentina, 100,000 in the Falkland Islands and 200,000 in Chile), and is classified as Near Threatened and in decline (ICMBIO, 2010; SILVA FILHO & RUOPPOLO, 2014; IUCN, 2016).

From September to March, Magellanic penguins are found in their breeding colonies. Two eggs are laid, which remain under the parental care of both parents for a period of 36 to 44 days. After the incubation period, the chicks remain under parental care for 60 to 120 days, after which they become independent. In February, the now young individuals go through the process of molting their feathers, an important event that precedes the pelagic life they will have in the non-reproductive period. At the end of April, they migrate in search of food in warmer waters, returning to the breeding colonies in September.

Migratory movements are probably influenced by the Falklands/Malvinas current (HOYO et al., 1992; SICK, 1997; VOOREN & BRUSQUE, 1999; PUTZ et al., 2000, 2007). The cold waters brought by this current represent a highly productive biomass system (CIOTTI et al., 1995). Thus, it is believed that the subtropical convergence determines the migratory pattern of these birds on the Brazilian coast, since it is likely that migratory movements are shaped by food distribution (PUTZ et al. 2000, 2007; ICMBIO, 2010). The migratory period begins at the end of February, due to the abundance of their prey. The birds remain on the high seas off the continental shelf of Argentina, Uruguay and Brazil during this period (SILVA FILHO & RUOPPOLO, 2014). Records of these birds are common on the coasts of Rio Grande do Sul and Santa Catarina, and can reach the coasts of the Southeast, such as Rio de Janeiro and Sao Paulo. Although rare, there are records of Magellanic penguins on the coasts of Bahia, Alagoas and Pernambuco (ICMBIO, 2010).

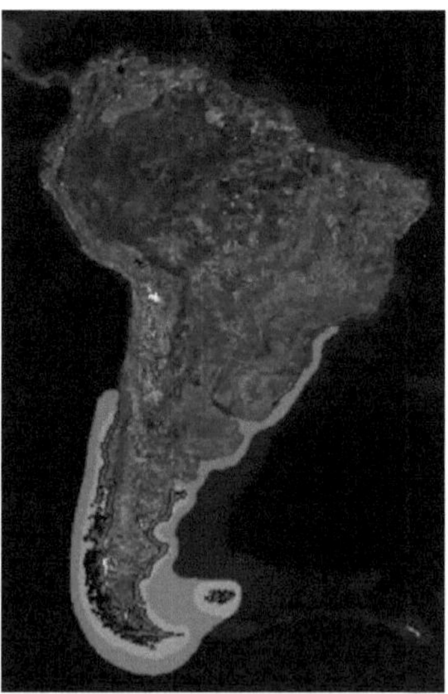

Figure 2 Distribution of Magellanic penguins in the Southern Hemisphere. I I Area

where penguins are usually found. I-I Non-breeding area for penguins, usually found during migration. I Area where breeding colonies are concentrated. Source: IUCN (2016)

2.3 Causes of Magellanic Penguin strandings

Although the world population of Magellanic penguins, with around 1.3 million breeding pairs, is not considered threatened, the two largest colonies in Argentina, especially the one in Punta Tombo, have been suffering from a sharp population decline in recent decades (IUCN, 2016). Habitat destruction, uncontrolled tourism, lack of food due to overfishing, marine pollution including oil spills, incidental capture in fishing and other factors are responsible for the population decline of this species (ICMBIO, 2010).

During their winter migration, these animals face various obstacles, many of which cause Magellanic penguins to strand in Brazil. The main factors influencing the weakening and mortality of these birds are oil contamination, parasites, fishing net accidents, ingestion of anthropogenic debris, among others. Mader et al (2010) carried out a survey from 1990 to 2008 on the coast of Rio Grande do Sul, and concluded that strandings start in June and peak in November. According to these authors, it is estimated that 19,500 penguins are found dead every year on the southern coast of Brazil, of which approximately 7,000 deaths are caused by anthropogenic and climatic factors.

2.3.1 Difficulty in obtaining food and accidental fishing

The difficulty in obtaining food during the migration of these birds is considered to be one of the main factors associated with the stranding and mortality of Magellanic penguins on the Brazilian coast (GEEVERGHESE, 2013). Their debilitation begins due to fleeing predators and intraspecific and interspecific competition for resources. As the migratory movements of these birds are shaped by the distribution of food, penguins follow their main prey, the anchovy (*Engraulis anchoita*) on its route through the cold currents to the Brazilian coast. However, penguins compete with the anchovy *(Pomatomus saltator),* a larger pelagic fish that approaches the coast in the colder months, when the anchovy shoals form (MADER, 2011).

In addition to this competition, the fish consumed by Magellanic penguins are commercialized species, which leads to a lack of food due to commercial overfishing. Another problem is the accidental capture of penguins in trawls. According to Cardoso et al. (2011), a single fishing trip with gillnets off the coast of Rio Grande do Sul resulted in the accidental capture of dozens of individuals. Other studies on the Pacific coast have shown that in some cases more than 1,300 animals can die in one catch (SIMEONE et al., 1999; MADER, 2011).

2.3.2 Ocean pollution and oil contamination

Marine pollution is a significant impact of human actions on the oceans. The dumping of domestic and industrial sewage, chemicals, garbage and other sources of pollution are increasingly common and in large quantities (GREENPEACE, 2016). Although all types of pollution affect the marine ecosystem, oil spills and garbage, especially plastics, are the most worrying when it comes to seabirds, especially penguins (WELTE & FRINK, 1991; BRANDAO et al., 2011).

Among the garbage deposited in the marine environment, plastics are the main threat to birds. They can be ingested voluntarily, when the bird mistakes them for prey, or involuntarily, when the plastic is already inside the prey (ROBARDS et al., 1995). When ingested, plastics can obstruct the animals' digestive systems and they are unable to eat any more, often dying from malnutrition. Although data for other penguins is scarce, litter ingestion in Magellanic penguins has increased over the years (AZEVEDO & SCHILLER, 1991; PINTO et al., 2007; MADER et al., 2010; TOURINHO et al., 2010). In a study carried out by Brandao et al. (2011), 175 dead penguins were collected in the winter of 2008 and 2010 in the Lagos region of Rio de Janeiro. Of these penguins, 15% had garbage in their stomachs, 57.69% of which was plastic, 42.31% items related to fishing and 7.69% other garbage.

Another worrying factor is the contamination of the oceans by oil. Since 1930, with the development of the oil industry, penguins have been negatively impacted by oil during their winter migration from Patagonia to northern Argentina, Uruguay and Brazil. This is because their migratory routes coincide with the maritime traffic of oil tankers and oil platforms (BOERSMA et al., 2011; GEEVERGHESE, 2013).

The increase in the number of Magellanic penguin carcasses found on the Brazilian coast is believed to be due to oil spills off the southern coast of Brazil and Uruguay. For example, in 1997, there were three oil spills on the beaches of Rio Grande do Sul; in June 2008, there was an oil spill caused by an accident between two ships off the coast of Uruguay; in September 2015, a diesel oil spill was recorded from an unidentified vessel by the Port Authority in Florianopolis, and associated with all these accidents, there was an increase in the number of penguin carcasses on the beaches of these locations (MADER et al., 2010; MADER, 2011).

Penguins are very vulnerable to oil spills compared to other seabirds, as they dive deep into the water, are unable to fly and cannot detect and avoid contamination (GARCIA-BORBOROGLU et al., 2006). When they become contaminated with oil (Figure 3), penguins lose their sense of humor.

This compromises their thermoregulation system and can cause hypothermia, forcing them to leave the water. When they leave the water, they lose their source of food, causing malnutrition and dehydration, as they take their food from the water. In addition, they can suffer internal damage, as they can ingest oil in an attempt to clean their feathers. The most common damage is intestinal irritation, which can increase dehydration and metabolic imbalance due to lack of food, malfunctioning of the liver and kidneys, inability to absorb nutrients or regulate fluids and electrolytes, which can cause bleeding in the intestinal tract (WELTE & FRINK, 1991).

Figure 3 Magellanic penguins contaminated with oil. (A) Penguins in a rehabilitation center, (B) Petrolized carcass found in Rio Grande do Sul in 2008. Source: Amor Animal (2015) and Mader (2011).

2.3.3 Climatic variations

The El Nino phenomenon influences the populations of the Galapagos penguin *(S. mendiculus)* and the Magellanic penguin in Argentina, causing increased mortality of individuals after major storms (MADER et al., 2010). El Nino can change the temperature of the water, making it colder or warmer, sea currents can change and major storms can occur. Warmer water can decrease the availability of food, lowering the immunity of these animals and, consequently, facilitating the proliferation of parasites in penguins, weakening them. Changes in sea currents can alter the course of penguins. In 2008, Magellanic penguins went further north than usual, almost reaching the Equator, in search of food. Some 3,371 penguins were recorded along 5,000 km of coastline. Many died, others were dehydrated, anemic, hypothermic and very thin due to lack of food (BOERSMA et al., 2011; MADER, 2011).

2.4 Rehabilitation centers

Marine animal rehabilitation centers are enterprises authorized by the Brazilian Institute of the Environment and Renewable Natural Resources - IBAMA, with the main objective of receiving, sorting and rehabilitating marine species so that they can be reintroduced into their natural habitat (IBAMA, 2008). Around 25 rehabilitation centers are located along the west and east coasts of the Atlantic, from northern Brazil to central Argentina. Various animals are treated in these centers, but approximately 64% of the seabirds taken are Magellanic penguins contaminated with oil (GARCIA-BORBOROGLU et al., 2006; BOERSMA et al., 2011).

2.4.1 Structure of the facilities

The management of Magellanic penguins must follow Normative Instruction No. 4, of March 4, 2002, of the Brazilian Institute for the Environment and Renewable Natural Resources (IBAMA, 2002). These birds are gregarious, so they should not be kept solitary. Environmental conditions are very important for the successful rehabilitation of these animals, and basically four requirements are needed to achieve this goal: a dry area, a swimming pool, ventilation and a quarantine area (AZA, 2005; SILVA FILHO & RUOPPOLO, 2014).

Magellanic penguins are able to tolerate environmental variations, with temperatures ranging from 0°C to 30°C. However, their comfort temperature ranges from 15°C to 20°C. Therefore, these birds must have a shaded area and a swimming pool so that their bodies can thermoregulate properly (SILVA FILHO & RUOPPOLO, 2014). Both salt and fresh water can be used in the pool. However, according to the study by Reisfeld et al. (2012), Magellanic penguins spent more time in pools with salt water than fresh water. Thus, the rates of pododermatitis and feather wear problems decreased, promoting an environment similar to the natural one. The pool should be easily accessible, preferably with a ramp to help the penguins get in and out of the water and a minimum depth of 0.6 meters (Figure 4). Penguins often defecate in the water, and although much of the feces sinks into the tank, the greasier portion floats, creating an oily film on the surface of the water, which contaminates the feathers and impairs waterproofing. To avoid this, it is necessary to have a drainage system that maintains continuous water circulation, with daily changes or the use of direct pumping of water from the estuary or the sea (AZA, 2005; SILVA FILHO & RUOPPOLO, 2014).

Figure 4 Example of a pool that allows penguins to enter and exit. Source: Silva Filho & Ruoppolo (2014).

In addition, a covered area is needed, protected from the cold, wind and rain, with conditions such that the birds can avoid excessive heat. Ideally, they should be exposed to natural light, so that they can follow the natural photoperiod and guarantee the production of vitamin D3 when exposed to the sun. The size of the enclosures should be in accordance with their occupancy density, in the case of Magellanic penguins, 2 individuals/8m^2 (IBAMA, 2002). The floors of the enclosures should be easy to sanitize with hoses, and can be made of burnt cement or granite, or even covered with newspapers or carpets (Figure 5A). Another option is to use rounded stones of approximately 3 to 7 cm, which reduce abrasion on the animals' feet as they help them to support their feet in different positions (Figure 5B). The enclosures should be very well ventilated, as penguins are sensitive to aspergillosis (SILVA FILHO & RUOPPOLO, 2014). As a form of prevention, it is necessary to have a quarantine area so that penguins suspected of having a disease, especially aspergillosis, do not contaminate other healthy penguins (AZA, 2005).

Figure 5 Flooring in penguin enclosures to prevent abrasion of the animals' feet. (A) Flooring covered with newspapers, (B) Flooring covered with stones. Source: Silva Filho & Ruoppolo (2014) and IPRAM (2016).

2.4.2 Rehabilitation of Magellanic penguins

The minimum requirements for rehabilitating any animal depend not only on the facilities used in the process, but on all the aspects involved in the work. Quick rescue, a prepared team and effective stabilization of the animal are essential for successful rehabilitation. The steps in the rehabilitation process vary from individual to individual, depending mainly on the species and condition of the animal. Although there is a need to change some of the steps, it is important that the rehabilitator tries to follow the protocol steps in all patients (MILLER, 2012; GEEVERGHESE, 2013).

The penguin rehabilitation protocol described below was based on information from Miller (2012) and Silva Filho & Ruoppolo (2014). It can be divided into 3 parts: admission of the animal, physical examinations and treatment (see Appendix A for a more detailed rehabilitation protocol).

When the penguins are admitted, the conditions and location in which each animal was found should be recorded and they should be identified with temporary rings (plastic seals with specific numbers for each individual). Depending on the condition in which the penguin arrives at the center, the veterinarian has some idea of what to do first, the most important step being to stabilize the animal. This stabilization consists of keeping the animal in a dry, ventilated place, away from mosquitoes and noises that could further stress the bird, and keeping it in a warm environment with infrared or 200W lamps.

The next step is to hydrate the animal, orally or subcutaneously at a warm temperature of 36°C to 39°C, to reduce regurgitation and avoid hypothermia. If the penguin is able to keep its head up, hydration is done orally and the dosage is 60 ml of saline solution every 1 hour and 30 minutes for the first 4 to 6 hours. After this period, the penguin is rehydrated three times a day. If the penguin is extremely weak or cannot hold its head up, hydration is done subcutaneously, with 20 ml/kg of saline solution once or twice a day. The initial feed is a paste containing a vitamin supply to make up for the lack of B vitamins that are lost by freezing the fish and can be 60 or 120 ml per day. As the penguin's health improves, it is fed fish slices, fish fillets and, finally, the whole fish. Both feeding and hydration are applied using gastric probing, with 60 ml syringes and a latex probe (about 30 cm long and 7 mm thick). The probe is inserted into the esophagus and the contents are released into the animal's stomach by restraining it. To restrain the animal, one person holds the individual between the knees and the ground, opens the beak with one hand and inserts the probe into the penguin's esophagus with the other.

Once the animal has been stabilized, the next step is a physical examination including weighing, blood and faeces collection, assessment of the colour of the mucous membranes, joints, temperature, percentage of petrolization, presence of wounds and body condition. If the animal shows blood in its feces, a solution of bismuth subsalicylate (2 ml/kg) is recommended twice a day until the clinical signs disappear. The penguin's normal hematocrit should be between 45-55%, the percentage of leukocytes should be equal to 1% and the total protein should be around $6g.dL^{-1}$. A hematocrit of less than 30% should be treated with hydrogenated iron dextran (10 mg/kg, intramuscularly). However, the application of medication and dosages are specific to each animal. And to monitor the animals' progress, weighing and blood sampling should be carried out weekly.

The penguin needs aspergillosis prevention. For this, itraconazole (20 mg/kg, orally, every 24 hours) is administered for 15 days or for as long as the animal remains in an environment prone to contamination with the fungus. If the animal has a parasitic disease, the antiparasitic drug should be administered when the penguin is no longer debilitated or dehydrated. In the case of oiled penguins, this protocol can be followed, adding a bath to remove the oil a few hours after the animal has stabilized. Over time, penguins that show improvement can be taken back to the pool to help with their recovery and subsequent reintroduction into the natural habitat.

2.4.3 Doengas in captivity

2.4.3.1 Viral diseases

The viral diseases already reported in Magellanic penguins are those caused by avipoxvirus and herpesvirus. *Avipox* belongs to the *Poxviridae* family and causes the disease known as yaws, variola or avian poxvirus. Transmission can occur directly, through contact with viral particles or with the damaged skin and mucous membranes of an infected individual; and indirectly through a vector, such as insect bites. There is no known specific treatment for penguins infected with avipoxvirus. The best method is to prevent the disease (TRIPATHY et al., 2008).

Herpesviruses, which belong to the *Herpesviridae* family, can be transmitted from bird to bird both by inhaling viral particles located in the feathers and by contact with saliva, urine, feces and secretions. The infected penguin can have hemorrhagic and necrotizing lesions that can spread to other tissues, such as lymphatic, epithelial and nervous (NIEMEYER, 2014). Herpesviruses have already been reported as the cause of death of African penguins in captivity and until 2014 there were no similar reports for Magellanic penguins (SILVA FILHO & RUOPPOLO, 2014). However, Niemeyer (2014) discovered a new herpesvirus species capable of causing hemorrhagic tracheitis and death in Magellanic penguins. Once the penguin is infected with the virus, it remains inside it for life, and viral reactivation can occur when the animal is immunosuppressed.

2.4.3.2 Bacterial diseases

Magellanic penguins have a skin and gastrointestinal microbiota, such as *Escherichia coli, Alcaligenes faccalis, Paracolon* spp., *Bacillus* spp., *Pseudomonas* spp., *Staphylococcus* spp., *Streptococcus* spp., among others, which are normal and beneficial to birds. However, this microbiota can be harmful to penguins by causing ententes, respiratory and systemic infections if the animal has low immunity , eats contaminated food and if there is any damage to the gastrointestinal tract (BOCARDO et al., 2009; SILVA FILHO & RUOPPOLO, 2014).

The most common bacterial diseases in Magellanic penguins undergoing rehabilitation are mycoplasmosis and salmonellosis and can be diagnosed through hematological evaluation, which can be used to delineate the disease and its prognosis (CAMPOS et al., 2009). Mycoplasmas can cause problems in the penguins' respiratory, urogenital, nervous and joint systems, while salmonellosis can cause systemic or enteric infections, depending on the bird and the species of bacteria (RASO et al., 2009).

2.4.3.3 Fungal diseases

Aspergillosis is the most common fungal disease and the one that causes the most deaths in Magellanic penguins undergoing rehabilitation. Immunosuppression associated with the stress of rehabilitation, as well as humid and poorly ventilated environments make penguins susceptible to the fungus *Aspergillus* spp. Prophylaxis and disinfection of the environment are determining factors in the fight against *Aspergillus* spp. as demonstrated by Xavier et al (2007). Younger birds are the most susceptible to contamination with the fungus, as their immune systems are not yet fully mature (FISHER et al., 2012).

The symptoms of aspergillosis are mainly lethargy, weight loss, isolation from the group, difficulty breathing and breathing with the mouth open. Treatment is mainly with itraconazole and nebulizers, but their effectiveness is limited depending on the degree of the disease (BOCARDO et al., 2009; SILVA FILHO & RUOPPOLO, 2014).

2.4.3.4 Parasitic diseases

Among the parasitic diseases in captive Magellanic penguins, the most common is malaria. However, there are other parasites such as gastric nematodes, intestinal cestodes and coccidiosis, as well as lice, ticks, trematodes and others that are found in animals found on the Brazilian coast and aggravate the animal's state of debilitation (BOCARDO et al., 2009; SILVA FILHO & RUOPPOLO, 2014).

Avian malaria is caused by the protozoan *Plasmodium* spp. transmitted by *Anopheles* and *Culex* mosquitoes. It is an asymptomatic disease for most birds, but for penguins, which have not co-evolved with these parasites, it is highly pathogenic. Prevention against this disease is based on the use of mosquito screens, as well as ventilation, preventive medication and repellents (CABANA et al., 2007; SILVA FILHO & RUOPPOLO, 2014; VANSTREELS, 2014).

2.4.3.5 Pododermatitis

Pododermatitis, also known as 'bumblefoot', is the most common disease in Magellanic penguins undergoing rehabilitation and occurs due to the abrasion and friction of the penguins' feet on the floors of the centers. When they are in their natural habitat, penguins spend most of their time at sea, and can go months without returning to dry land. And when this happens, the soil is not very abrasive, such as stones, sand, foliage, among others, which makes it difficult for pododermatitis to form (SILVA FILHO & RUOPPOLO, 2014). The lesions caused by pododermatitis cause the proliferation of microorganisms in the subcutaneous tissue, which leads to the formation of cases. If left untreated, they can become infected and lead the penguin to septicemia and death (BOCARDO et al., 2009).

3 MATERIAL AND METHODS

3.1 Study Area

The study was carried out in September 2015 at Instituto Gremar - Pesquisa, Educação e Gestao de Fauna, a marine animal rehabilitation center located on the Guaruja-Bertioga road, in the municipality of Guaruja, São Paulo. Its area of operation includes the entire Baixada Santista (Figure 6), from Peruibe to Bertioga. Since 2004, the institute has been working on environmental monitoring, rehabilitation of victimized animals, environmental education activities and responding to environmental emergencies involving wildlife.

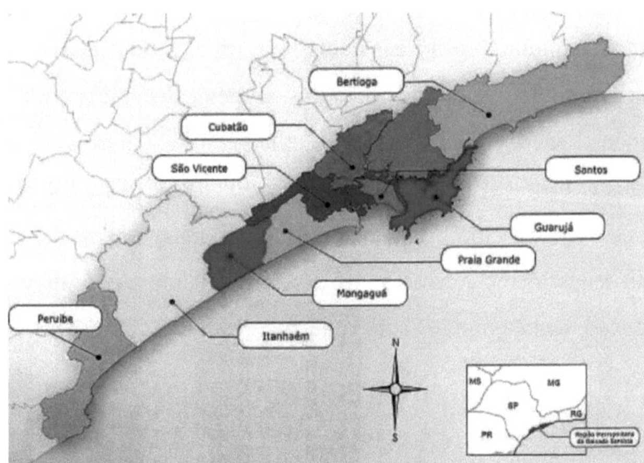

Figure 6 The Gremar Institute's area of operation in Baixada Santista. Source: Google (2016).

3.2 Management of Magellanic Penguins

The animals that arrive at the Institute are mainly taken in by the Environmental Police , who receive calls from the local community when they spot a stranded penguin. Approximately 120 penguins arrived at the Gremar Institute between June and August 2015. No stranded penguins arrived in September. All received care from the Institute's staff based on the protocol mentioned in section 2.4.2, with the exception of a few details. The drug itraconazole was not given to the penguins when they arrived and instead of mixing B vitamins into the fish paste, cod liver oil was used. Those who were more stable, who ate fillets or whole fish, were given a thymine tablet once a week.

The penguins at Gremar were handled every day to avoid contamination with *Aspergillus* spp. Water and chlorine were used to disinfect the facilities every morning and the water in the pools was changed every afternoon. Despite this, three penguins were in the clinic with suspected aspergillosis.

These individuals were isolated from the others and treated with the antibiotic amoxicillin twice a day, 12 hours apart, and then with itraconazole once a day, 24 hours apart. Nebulizer sessions were rarely carried out. In addition, 60 ml of saline solution was administered three times a day, as well as feeding whole fish.

The fully rehabilitated penguins followed the routine of staying in the pool for most of the day (Figure 7) and returning to the enclosure at night. They were fed twice a day with whole fish and 60 ml of saline solution in the morning. These penguins were now ready to return to their natural environment. Before being reintroduced to their natural *habitat*, they were given permanent metal rings, so that if they stranded again, the Institute would know that the penguin was a repeat offender.

Penguins with any signs of pododermatitis wore a bandage on their feet ("booties") which was changed whenever necessary, usually once a week. Once a month, the individuals received a dose of anti-parasites.

Figure 7 Rehabilitated penguins swimming in the pool at the Gremar Institute.

4 RESULTS AND DISCUSSION

Of the 120 young penguins that arrived at Gremar, only 25 remained in September. Of these, three were suspected of having aspergillosis, one was a chick and the other 21 were in favorable conditions for release due to their complete rehabilitation. The high mortality of penguins in rehabilitation centers is common. In a study carried out by Pereira et al. (2012), of 203 penguins that arrived at the rehabilitation center of the Federal University of Rio Grande - FURG, 132 died and only 24 were released after rehabilitation. In another study carried out between 1993 and 1995, only 40% of the penguins rescued were released in conditions favorable to survival (MICHELS-SOUZA et al., 2010). These results indicate that despite efforts to rehabilitate Magellanic penguins in rehabilitation centers, new preventive measures must be implemented to avoid stranding these animals.

No viral or bacterial diseases were recorded in the penguins at the Gremar Institute. However, aspergillosis was suspected in three individuals (P593, P574 and P335). At the beginning of September, penguins P593 and P335 were isolated from the others and treated with serum and fish three times a day, the first serum containing aminomix. They were also given amoxicillin at 8am and 8pm. In the same week, individual P593 did not resist treatment and died. Individual P574, who appeared to be in good health with the other penguins, showed symptoms of weakness and difficulty standing. He was placed in isolation with suspected aspergillosis and received the same treatment as the other two penguins. Failing to respond to treatment after two days, P574 was given a dose of itraconazole, but the following day he was also dead . Individual P335 was the one who resisted the disease the most. Some nebulization sessions were carried out on this penguin, which showed a slight improvement. However, the sessions were interrupted because the nebulizer was damaged, his condition worsened again, a dose of itraconazole was administered, but the next day he died.

Aspergillosis can be classified as acute and chronic. Generally, the acute phase occurs when the individual is exposed to large quantities of spores, progressing rapidly and leading to death within 48 hours. In most cases, it is severe, causing dyspnea with a high mortality rate (XAVIER et al., 2007). Penguin P574 showed symptoms of the disease overnight, and despite treatment, his death occurred approximately two days later. It can therefore be considered a case of acute aspergillosis. The other two individuals had characteristics of chronic aspergillosis. This last type usually occurs in individuals who have been exposed to low levels of spores, but who are immunosuppressed due to stress and malnutrition (XAVIER et al., 2007).

Diagnosis of aspergillosis is difficult because the symptoms are not specific. Generally, the diagnosis depends on various tests such as serology, hematology, biochemistry, endoscopy, fungal culture and others. In addition, lethargy, weight loss, isolation, inappetence, loss of vocalization, open-mouth breathing, coughing and difficulty in standing up can help diagnose aspergillosis (AZA, 2005; BEERNAERT et al., 2010). Penguins P593 and P335 showed lethargy, open-mouthed breathing, weight loss and inappetence for a long time (Figure 8). On the other hand, penguin P574 showed signs of weakness and difficulty standing from one day to another.

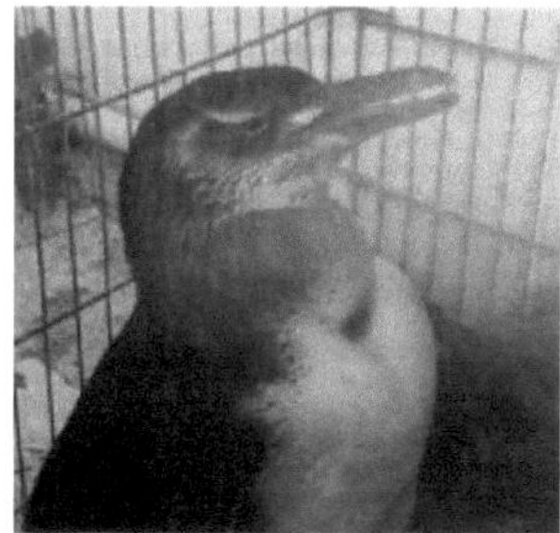

Figure 8 Penguin showing dyspnea due to aspergillosis. Note the open beak during breathing and the swelling of the region cranial to the keel. Source: Silva Filho & Ruoppolo (2014).

The method and success of treatment depend a lot on the individual and the stage of the disease. Generally, treatment for aspergillosis consists mainly of the use of nebulizers and medication with itraconazole, 10 to 15 mg/kg, 24 hours (AZA, 2005; SILVA FILHO & RUOPPOLO, 2014; CUBAS et al., 2014). However, the Institute's penguins were treated differently. The antibiotic amoxicillin was used constantly from the beginning of the symptoms and nebulization was only used on penguin P335. The veterinarians' explanation for using the antibiotic was that it was necessary to prevent the animal from being infected by any bacteria, since they were weakened by aspergillosis. However, as the treatment was only done with antibiotics, the penguins' condition only worsened, as the main cause, fungi, was not combated. It was therefore decided to administer a dose of itraconazole to each penguin. Itraconazole is a very strong drug and as the penguins were weak from the disease, they couldn't resist for long and died the day after it was administered. What could have been done from the outset was to supplement the animals' diet with vitamins and other nutrients, and along with the antibiotic, constant nebulization and application of itraconazole when they were strong. This would give them a better chance of fighting off the fungus and possibly resisting the disease.

The main ways of preventing the disease are: constant disinfection of penguin facilities, well-ventilated environments and the administration of antifungals to penguins susceptible to the disease. Despite this, in some cases preventive methods do not prevent the outbreak of the disease, which is still difficult to control (DIEBOLD et al., 1999; XAVIER et al., 2007; SILVA FILHO & RUOPPOLO, 2014). All the penguins were not given antifungals and the enclosure in which they spent the night did not have adequate ventilation. This may have facilitated the contamination of the individuals with the fungus responsible for aspergillosis.

As the symptoms and diagnosis of the disease are still not very specific, a necropsy was carried out in order to be sure whether or not aspergillosis was the cause of the penguin's death. The necropsy was carried out on the same day or the day after the penguins died. The cause of death of the three individuals was confirmed to be aspergillosis. In individuals P593 and P335, the disease was spread throughout the body, with fungal colonies and cases in organs such as the lung, liver, heart and other tissues. Individual P574 had colonies concentrated in the air sacs, the textures of which were thick and rigid due to the severe infection (Figure 9).

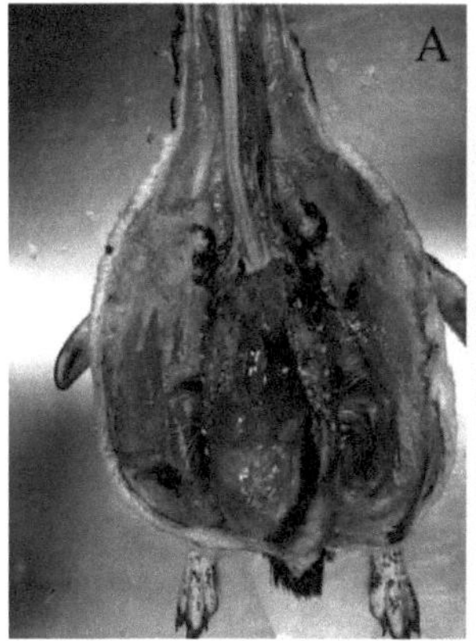

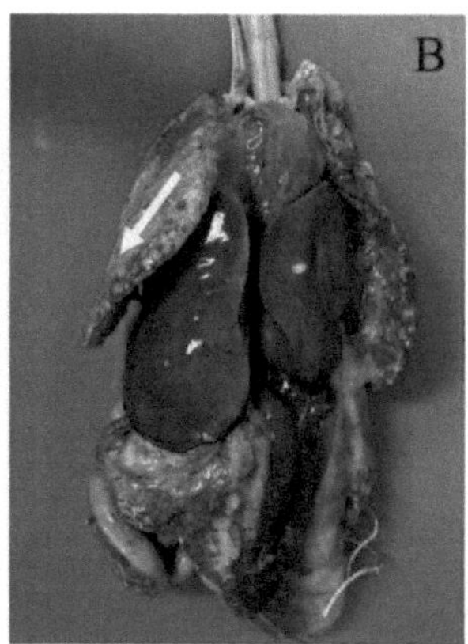

Figure 9 Necropsy of individual P574. (A) Beginning of the necropsy, (B) View of the adrenal sacs.
completely taken over by colonies of the aspergillosis fungus. Note the white arrow, indicating a colony in this individual's liver.

All the penguins at Gremar took anti-parasite medication once a month. Despite this, during the necropsy of penguins P593 and P335, some nematodes were found in their digestive tracts (Figure 10). Generally, penguins who have just arrived at a rehabilitation center have many parasites, which take advantage of the individuals' low immunity due to inanition from the long journey they face. The most common parasite found in Magellanic penguins is the nematode *Contracaecum pelagicum,* characteristic of piscivorous birds from the southern hemisphere (MADER, 2011; SILVA FILHO & RUOPPOLO, 2014). Therefore, it is likely that the parasites found during the necropsy of the two individuals was this nematode. However, a study is needed to identify the species with certainty. Although it is very common in Magellanic penguins, none of the penguins at the Institute have been diagnosed with malaria.

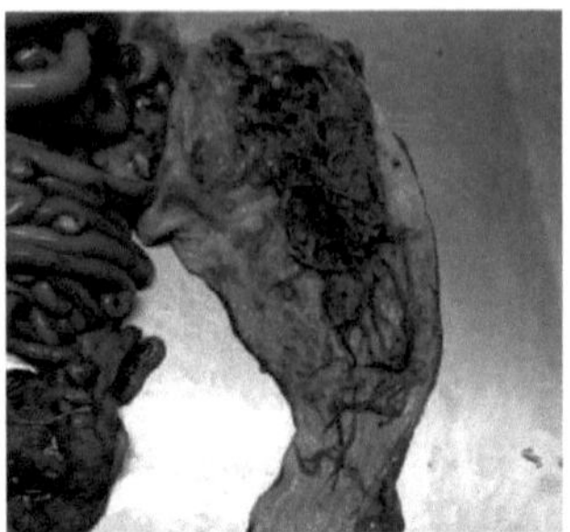

Figure 10 Parasites found in the digestive tract of individual P574.

Despite the constant cleaning of the enclosures and the use of rubber mats as flooring, all the penguins at Gremar had some level of pododermatitis (Figure 11). To control the infection, Hipoglos® ointment was applied to the wounds and the paws were wrapped with Vetrap® tape, gas and adhesive plaster. This reduced the chance of the disease getting worse, as it protected the paws from abrasion from the ground. Various other treatments can be used, such as foot baths with disinfectants, solutions containing dimethylsulfoxide, among others. The painful sensation caused by pododermatitis forces animals to stand on their tarsometatarsus instead of using their feet. This position prevents pain for a while, but causes other injuries. If the position persists, they will spend most of their time lying down, which will cause contact with the feces on the floor, rotting the feathers and exposing the skin of the keel (Figure 12) (SILVA FILHO & RUOPPOLO, 2014).

These treatments help to slow down the lesions, but none of them guarantee a complete cure. For this reason, the best way to deal with this disease is to place smooth rounded stones on the floor, as well as rubber mats or plastic , clean the floor daily and encourage the animal to swim as much as possible. Surgery should be used as a last resort, as post-surgery can lead to new injuries due to changes in the animal's posture (SILVA FILHO & RUOPPOLO, 2014).

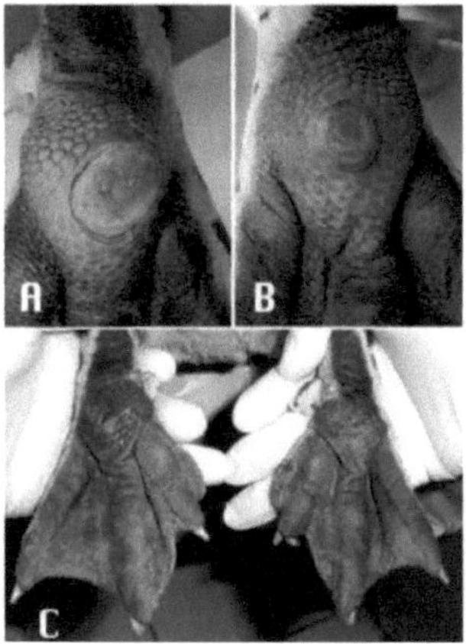

Figure 11 Evolution of pododermatitis in a Magellanic penguin. (A) advanced stage lesion, (B) healing lesion, (C) healed lesion. Source: Reisfeld et al (2013).

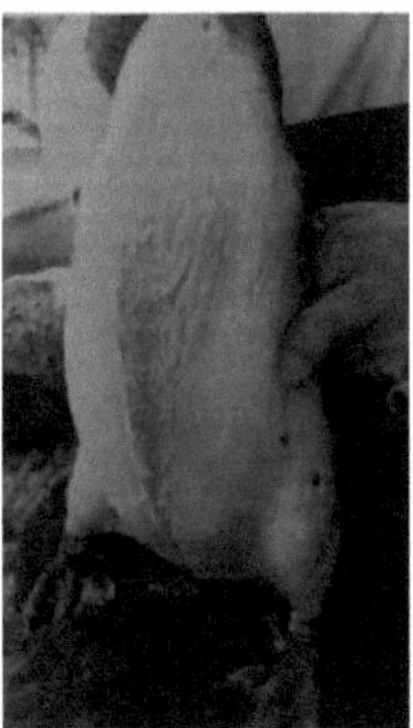

Figure 12 Feathers rotting on the ventral side of the individual as a result of body support. Source: Silva Filho & Ruoppolo (2014).

The baby penguin arrived at the Gremar very weak, unable to stand on its own. The rehabilitation protocol was applied and by September he was able to stand up and eat whole fish (Figure 13A). The other 21 penguins were released in the Laje de Santos Marine State Park on September 22 (Figure 13B). As these penguins were released in September, they return to their breeding colonies where they look for partners to reproduce with, continuing their life cycle.

Figure 13 (A) Penguin P144 ready for its first swim in the pool, (B) Rehabilitated penguins released on the Laje de Santos.

5 FINAL CONSIDERATIONS

The solution for stranded penguins is rehabilitation. However, this process requires a specialized team, time and financial resources, which is not always possible. In addition, most of the factors that lead to penguin strandings are anthropogenic causes. Only a fraction of the penguins affected by any cause reach the beaches alive and many of these do not survive for long if they are not treated immediately.

When the penguins arrive at the Gremar Institute, the main concern is whether or not they are ill, as well as preventing the appearance of diseases in individuals already at the center. As shown, aspergillosis is a determining factor in the rehabilitation of Magellanic penguins. Therefore, extreme care must be taken with the hygiene of the facility.

Another important issue is the application of Gremar's rehabilitation protocol. Some changes have been made by the Institute, but they haven't worked. The three penguins suspected of having aspergillosis did not resist the disease and were basically treated with antibiotics. It is therefore necessary for Gremar's veterinarians to meet to study a more efficient approach to combating this disease.

Penguins are "flagship species", which means that, because they are charismatic, they can be used to adopt strategies and environmental awareness programs to preserve their natural habitat and, consequently, all the other species that share the same environment. To this end, stricter measures are needed to curb the decline of the penguin population, such as stricter penalties for oil companies, the fishing industry and anyone else, whether physical or legal, who disrespects conservation rules.

BIBLIOGRAPHICAL REFERENCES

ANIMAL LOVE. Penguin migration period in Brazil. Available at: http://amor-animal.org/periodo-migratorio-dos-pinguins-no-brasil/. Accessed on: 02 Feb. 2016.

AZA - American Zoo and Aquarium Association. **Penguin husbandry manual.** Chapter 3: Behavior and social organization. 2005. 142 p.

AZEVEDO, T.R., SCHILLER, A. Notes on the diet and the ingestion of plastic material by the Magellanic penguin *Spheniscus magellanicus* on Santa Catarina Island and mainland (Brazil). **Research Report**, University of Liege, Institution of Zoology, Belgium. Rapport, v. 457, p. 1-8, 1991.

BEERNAERT, L.A et al. Aspergillus infections in birds: a review. **Avian Pathology**. v. 39, n. 5, p. 325-331, 2010.

BOCARDO, M.; MACHADO, J.A.C.; PEREIRA, R.E.P. Main diseases in penguins - literature review. **Electronic Scientific Journal of Veterinary Medicine**. v. 12. 2009.

BOERSMA, P.D.; GARCIA-BORBOROGLU, P.; RUOPPOLO, V. International Society for the Conservation of Penguins. In: **Bulletin N° 01 - Penguins in Brazil**. 2011. p. 3-4.

BRANDAO, M.L.; BRAGA, K.M.; LUQUE, J.L. Marine debris ingestion by Magellanic penguins , *Spheniscus magellanicus* (Aves: Sphenisciformes), from the Brazilian coastal zone. **Marine Pollution Bulletin**. v. 62, p. 2246-2249, 2011.

CABANA, A.L. et al. Malaria in a *Magellanic* penguin *(Spheniscus magellanicus)* - Case report. In: SALAO DE 1NICIACAO CIENTIFICA, 19. 2007. Porto Alegre, RS. **Book of abstracts**. Porto Alegre, UFRGS, 2007.

CARDOSO, L.G. et al. Gillnet fisheries as a major mortality factor of Magellanic penguins in wintering areas. **Marine Pollution Bulletin.** v. 62, n. 4, p. 840-844, 2011.

CAMPOS, S.D.E. et al. Hematological evaluation of nine *Magellanic* penguins *(Spheniscus magellanicus* Foster, 1781) kept in captivity. **Veterinaria em Foco**, Canoas. v. 6, n. 2, p. 165-171, 2009.

CIOTTI, A.M. et al. Freshwater outflow and subtropical convergence influence on phytoplankton biomass on the southern Brazilian continental shelf. **Continental Shelf Research**. v. 15, n. 14, p. 1737-1756, 1995.

CUBAS, Z.S.; SILVA, J.C.R.; CATAO-DIAZ, J.L. **Tratado de Animais Silvestres: Medicina Veterinaria Volume 1**. 2ª ed. Sao Paulo: Roca, 2014. 2470 p.

DIEBOLD, E.N.; BRANCH, S.; HENRY, L. Management of penguin populations in North American zoos and aquariums. **Marine Ornithology**. v. 27, p. 171-176, 1999.

FISHER, P.F.; SOUZA, J.; BERSELLI, M. Avian aspergillosis - literature review. In: XVII SEMINARIO INTERINSTITUCIONAL DE ENSINO, PESQUISA E EXTENSAO. 2012. Cruz Alta, RS. **Proceedings**. Cruz Alta, UNICRUZ, 2012.

GARCIA-BORBOROGLU, P. et al. Chronic oil pollution harms Magellanic penguins in the Southwest Atlantic. **Marine Pollution Bulletin**. v. 52, p.193-198, 2006.

GEEVERGHESE, C. **Rehabilitation of *Magellanic* penguins *(Spheniscus magellanicus)* shipwrecked on the beaches of the Brazilian coast: a literature review**. 2013. 35 p. Monograph (Graduation in Veterinary Medicine) - University of Brasilia, Brasilia, 2013.

GREENPEACE. Pollution. Available at: http://www.greenpeace.org/portugal/pt/O-what-we-do/oceans/pollution/. Accessed on: 15 Jan. 2016.

HOYO, J.D.; ELLIOTT, A. and SARGATAL, J. **Handbook of the birds of the world: Ostrich to Ducks**. Barcelona: Lynx Editions, 1992. 696 p. Volume 1.

IBAMA - Brazilian Institute for the Environment and Natural Resources Renewable. **Normative Instruction No. 04** of March 4, 2002.

IBAMA - Brazilian Institute for the Environment and Renewable Natural Resources. **Ibama Normative Instruction IN 169/08 & amendments of** February 20, 2008. Authorization for Wild Fauna Enterprises. 2008.

ICMBIO. **National Monitoring Project for the Magellanic Penguin *(Spheniscus magellanicus)*,** 2010.

GREMAR INSTITUTE - RESEARCH, EDUCATION AND WILDLIFE MANAGEMENT. Available at: http://www.gremar.org.br/index.php. Accessed on: Feb. 2, 2016.

IPRAM - Institute for Research and Rehabilitation of Marine Animals. **Penguins**. Available at: http://ipram-es.blogspot.com.br/p/pinguins.html. Accessed on: Feb. 2, 2016.

IUCN - INTERNATIONAL UNION FOR CONSERVATION OF NATURE. ***Spheniscus magellanicus*,** 2016.

MADER, A.; SANDER, M.; CASA JR., G.; Seasonal mortality cycle of the *Magellanic* penguin, *Spheniscus magellanicus*, influenced by anthropic and climatic factors on the coast of Rio Grande do Sul, Brazil. **Revista Brasileira de Ornitologia**. v. 18, n. 3, p. 228-233, 2010.

MADER, A. Why are so many Magellanic penguins dying in Brazil? In: **Bulletin N° 01 - Penguins in Brazil**. 2011. p. 6-7.

MICHELS-SOUZA, M.A. et al. Oil-contaminated *Magellanic* penguins *(Spheniscus magellanicus* Foster, 1781) found on the island of Santa Catarina, Brazil.
UEPG: Biological and Health Sciences. v. 16, n. 1, p. 29-34, 2010.

MILLER, E.A. **Minimum Standards for Wildlife Rehabilitation 4th edition.** National Wildlife Rehabilitators Association. St. Cloud, MN. 2012. 116 p.

NIEMEYER, C. **Investigation of selected viral diseases in seabirds from the Atlantic Coast of South America**. 2014. 117 f. Thesis (Doctorate in Sciences) - Faculty of Veterinary Medicine and Zootechny, University of Sao Paulo, Sao Paulo, 2014.

NOGUEIRA, N.F.; CORREA, C.G.; FEDULLO, L.P.L.. Reintroduction of the *Magellanic* Penguin *(Spheniscus magellanicus* Forster, 1781) after rehabilitation procedures. In: CONGRESSO DA SOCIEDADE DE ZOOLOGICOS E AQUARIOS DO BRASIL, 39., 2015. Foz do Iguagu. **Abstracts**. Foz do Iguagu: 39th Congress of the Zoological and Aquarium Society of Brazil, 2015.

PENGUINWORLD. *Spheniscus.* Available at: http://www.penguinworld.com/types/spheniscus.html. Accessed on: January 13, 2016.

PEREIRA, A. et al. Overview of *Magellanic* penguins *(Spheniscus magellanicus)* received in the rehabilitation sector of CECLIMAR/IB/UFRGS between 2009 and 2011. In: CONGRESSO LATINO-AMERICANO DE REABILITAGAO DE FAUNA MARINHA, 2. 2012. Rio Grande. **Abstracts**. Rio Grande: 2nd Latin American
American Marine Fauna Rehabilitation, 2012.

PINTO, M.B.L.C., SICILIANO, S., DI BENEDITTO, A.P.M. Stomach contents of the Magellanic penguin Spheniscus magellanicus from the northern distribution limit on the Atlantic coast of Brazil. **Marine Ornithology**. v. 35, p. 77-78, 2007.

PUTZ, K.; INGHAM, R.J. and SMITH, J.G. Satellite tracking of the winter migration of Magellanic penguins (Spheniscus magellanicus) breeding in the Falkland Islands. **Ibis**: a quarterly journal of ornithology, v. 142, p. 614-622, 2000.

PUTZ, K.; INGHAM, R.J. and SMITH, J.G. Winter migration of Magellanic penguins (Spheniscus magellanicus) from the southernmost distributional range. **Marine Biology**, Berlin: DE, v. 152, p. 1227-1235, 2007.

RASO, T.F. et al. Investigation of *anti-Mycoplasma synoviae, Mycoplasma gallisepticum, Salmonellapullorum* and *Salmonella gallinarum* antibodies in *Magellanic* penguins *(Spheniscus magellanicus).* **Biologico**, Sao Paulo. v. 71, n. 2, p. 83-202, 2009.

REISFELD, L. et al. Behavioral Responses of *Magellanic* Penguins *(Spheniscus magellanicus)* to freshwater vs. saltwater and their importance for captivity. In: CONGRESSO LATINO-AMERICANO DE REABILITAGAO DE FAUNA

MARINHA -RIO GRANDE DO SUL, BRASIL, 2. 2012. Rio Grande do Sul. **Abstracts**. Rio Grande do Sul: 2nd Latin American Congress on Marine Fauna Rehabilitation, 2012.

REISFELD, L. et al. Reducing bumblefoot lesions in a group of captive Magellanic penguins (*Spheniscus magellanicus*) with the use of environmental enrichment. **Pesquisa Veterinaria Brasileira**. v. 33, n. 6, p. 791-795, 2013.

ROBARDS, M.D., PIATT, J.F., WOHL, K.D. Increasing frequency of plastic particles ingested by seabirds in the subarctic North Pacific. **Marine Pollution Bulletin**. v. 30, p. 151-157, 1995.

SANDER, M. A little history about the Magellanic Penguin. In: **Bulletin N° 01 - Penguins in Brazil**. 2011. p. 2.

SICK, H. 1997. **Brazilian ornithology: an introduction**. 1 ed. Rio de Janeiro, Nova Fronteira, 912 p.

SILVA FILHO, R.P.; RUOPPOLO, V. **Tratado de Animais Selvagens: Medicina Veterinaria.** Chap.7: Sphenisciformes (Penguins). Ed. Roca. Sao Paulo, Brazil, 2014, 2512 p.

SIMEONE, A.; BERNAL, M. and MEZA, J. Incidental mortality of Humboldt Penguins Spheniscus humboldti in gill nets, Central Chile. **Marine Ornithology**. v. 27, p. 157161, 1999.

SIWANOWICK, I. *Animals in Close-up:* See the world's most incredible creatures up close. Sao Paulo: Nobel, 2010. 96p.

TOURINHO, P.S., IVAR DO SUL, J.A., FILLMANN, G. Is marine debris ingestion still a problem for the coastal marine biota of southern Brazil? **Marine Pollution Bulletin**. v. 60, p. 396-401, 2010.

TRIPATHY, D.N.; REED, W.M. Pox. In: SAIF, Y.M. **Diseases of poultry**. Oxford: Blackwell Publishing, 2008. p. 291-307.

VANSTREELS, R.E.T.D.V.O. **Study of avian malaria and other hemoparasites**

in penguins on the Atlantic coast of South America. 2014. Thesis (PhD in Experimental and Comparative Pathology) - Faculty of Veterinary Medicine and Zootechny, University of Sao Paulo, Sao Paulo, 2014.

VOOREN, C.M. and BRUSQUE, L.F. **The birds of Brazil's coastal environment: biodiversity and conservation**. Rio Grande, University of Rio Grande Foundation (Pronabio). 1999.

WALKER, B.G.; BOERSMA, P.D. **Diving Behavior of Magellanic Penguins *(Spheniscus magellanicus)* at Punta Tombo.** Canadian Journal of Zoology. 2003.

WELTE, S.; FRINK, L. Rescue and Rehabilitation of Oiled Birds. In. FREDRICKSON, L.H. **Waterfowl management handbook (Fish and wildlife leaflet)**. Washington: U.S. Department of the Interior, Fish and Wildlife Service, 1991. p.1-8.

WILLIAMS, T.D. 1995. **The Penguins**. Oxford: Oxford University Press.

XAVIER, M.O et al. Aspergillosis: A limiting factor during recovery of captive Magellanic penguins. **Brazilian Journal of Microbiology**. v. 38, p. 480-484, 2007.

ANNEXES

Annex A: Chronological outline of the rehabilitation process - Minimum Care Procedures for Wildlife Rehabilitation

1) Admission of the Animal

a) Gather history from the person presenting the animal

b) Record all information (Forms 4 &5, Appendix A)

c) Provide relevant educational material to the presenter

d) For baby animals, determine if the animal is in need of rehabilitative care or could be placed back in Its nest/den; if the latter, arrange return of the animal(s) accordingly (Charts 1 &2, Appendix B)

2) Stabilization of the Animal

a) Evaluate the animal quickly when transferring to a holding pen/cage/etc.

b) Examine for critical conditions and administer emergency care as needed

c) Provide warmth (unless hyperthermic)

d) Provide quiet rest space

e) Prepare materials needed for exam

Before handling the animal for the full examination, observe quietly for posture and behavior

3) Initial Examination

a) Animal identification (species, age, gender if sexually dimorphic)

b) Weight

c) Temperature (as able)
d) Visual exam
e) Palpate limbs
f) Examine holes
g) Assess nutritional status and condition

4) Initial Treatment
a) Provide fluids
b) Clean and treat any wounds
c) Stabilize fractures
d) Administer medications (antibiotics, analgesics, remedies, etc.)
e) Provide appropriate, palatable nutrition for species and status
f) Conduct or schedule any ancillary diagnostics (radiographs, bloodwork, fecals, etc.) and any additional treatments (surgeries, follow-up wrap changes, etc.)

5) Intensive Rehabilitation
a) Monitor weight
b) Provide ongoing, appropriate fluids and nutrition
c) Treat medical problems as needed
d) Provide comfortable, appropriate housing and habitat minimize interaction with human activity

6) Intermediate Rehabilitation (restricted activity)
a) Monitor weight
b) Provide ongoing, appropriate nutrition
c) Treat medical problems as needed (should be minimal)
d) Provide comfortable, appropriate housing and habitat with

mental stimulation, minimize interaction with human activity
e) Provide manual physical therapy as needed
f) Monitor behavior

7) Pre-Release Conditioning (unlimited activity)
a) Provide larger, outdoor housing
b) Monitor weight and general condition
c) Provide ongoing, appropriate nutrition, introducing a more natural diet
d) Treat any primary or secondary medical problems as needed (should be minimal)
e) Exercise daily, as appropriate for that species

8) Release Evaluation (exceptions for some of the following items may occur when natural circumstances cannot be replicated in captivity)
a) Ability to self-feed (perhaps catch live prey)
b) Normal mobility and function, reasonable level of physical fitness and stamina necessary for foraging, breeding, migration, or territory defense behavior if predicted
c) No evidence of disease
d) Normal weight for that species/sex/season
e) Normal blood values (where appropriate/feasible and known)
f) Suitable release sites available (see Section 7.2)
g) Normal behavior (the animal exhibits reasonable responses to human activity, exhibits normal socialization with both same and other species)

9) Release
a) Provide proper/safe transportation

b) Choose appropriate time of year (migration, breeding season, etc.)
c) Choose appropriate time of day
d) Provide food if appropriate
e) Monitor post-release activity and behavior, if possible

Printed by Books on Demand GmbH, Norderstedt / Germany